Gadgets

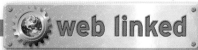
web linked

Gadgets

By Steve Parker

Illustrated by Alex Pang

First published in 2009 by Miles Kelly Publishing Ltd
Bardfield Centre, Great Bardfield, Essex, CM7 4SL

MASON CREST PUBLISHERS INC.
370 Reed Road, Broomall, Pennsylvania 19008
(866)MCP-BOOK (toll free), www.masoncrest.com

First Printing
9 8 7 6 5 4 3 2 1

Library of Congress Cataloging-in-Publication Data
Parker, Steve, 1952–
Gadgets / by Steve Parker ; illustrated by Alex Pang.
p. cm. — (How it works)
Includes bibliographical references and index.
ISBN 978-1-4222-1795-5
Series ISBN (10 titles): 978-1-4222-1790-0
1. Household electronics—Juvenile literature.   2. Miniature
electronic equipment—Juvenile literature. I. Pang, Alex, ill. II.
Title.
TK7880.P37 2011
621.3—dc22
                    2010033617
Printed in the U.S.A.

First published by Miles Kelly Publishing Ltd
Bardfield Centre, Great Bardfield, Essex, CM7 4SL
© 2009 Miles Kelly Publishing Ltd

Editorial Director: *Belinda Gallagher*
Art Director: *Jo Brewer*
Design Concept: *Simon Lee*
Volume Design: *Rocket Design*
Cover Designer: *Simon Lee*
Indexer: *Gill Lee*
Production Manager: *Elizabeth Brunwin*
Reprographics: *Stephan Davis, Ian Paulyn*
Consultants: *John and Sue Becklake*

Every effort has been made to acknowledge the source and
copyright holder of each picture. The publisher apologizes for any
unintentional errors or omissions.

WWW.FACTSFORPROJECTS.COM

ACKNOWLEDGMENTS

All panel artworks by Rocket Design
The publishers would like to thank the following
sources for the use of their photographs:
Alamy: 8 sciencephotos
Corbis: 6(t/r) Carl & Ann Purcell; 7(r) Fred Prouser/
Reuters; 14 TScI/NASA/Roger Ressmeyer; 16 Reuters;
22 Ed Kashi; 32 Roger Ressmeyer
Fotolia: 6(c) Avava; 7(t) Henrik Andersen;
25 Monkey Business; 27 Peter Baxter; 35 Andy Dean
Rex Features: 28 Action Press; 30 c.W.Disney/Everett
Science Photo Library: 11 Chris Martin-Bahr;
20 James King-Holmes; 37 David Hay Jones
All other photographs are from Miles Kelly Archives

Each top right-hand page directs you to the Internet to help you find out more. You can log on to **www.factsforprojects.com** to find free pictures, additional information, videos, fun activities, and further web links. These are for your own personal use and should not be copied or distributed for any commercial or profit-related purpose.

If you do decide to use the Internet with your book, here's a list of what you'll need:
• A PC with Microsoft® Windows® XP or later versions, or a Macintosh with OS X or later, and 512Mb RAM

• A browser such as Microsoft® Internet Explorer 8, Firefox 3.X, or Safari 4.X
• Connection to the Internet via a modem (preferably 56Kbps) or a faster Broadband connection
• An account with an Internet Service Provider (ISP)
• A sound card for listening to sound files

Links won't work?
www.factsforprojects.com is regularly checked to make sure the links provide you with lots of information. Sometimes you may receive a message saying that a site is unavailable. If this happens, just try again later.

Stay safe!
When using the Internet, make sure you follow these guidelines:
• Ask a parent's or a guardian's permission before you log on.
• Never give out your personal details, such as your name, address, or e-mail.
• If a site asks you to log in or register by typing your name or e-mail address, speak to your parent or guardian first.
• If you do receive an e-mail from someone you don't know, tell an adult and do not reply to the message.
• Never arrange to meet anyone you have talked to on the Internet.

The publisher is not responsible for the accuracy or suitability of the information on any website other than its own. We recommend that children are supervised while on the Internet and that they do not use Internet chat rooms.

CONTENTS

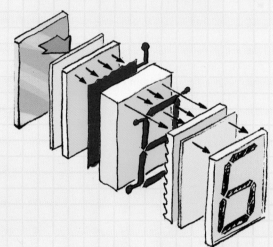

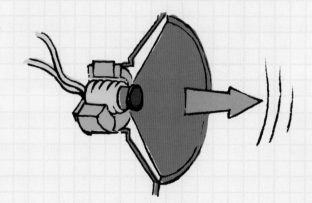

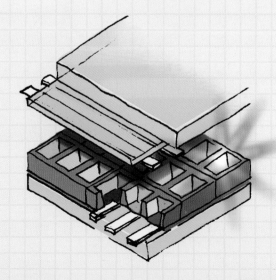

INTRODUCTION

People have always loved gadgets. In ancient times, alongside important tools and machines such as levers and wheels, people also invented gadgets, gizmos, widgets, and devices to make life easier and more fun. Many gadgets were novelties that came and went, used just for entertainment. But others took on more serious uses and became part of everyday life, such as the abacus, alarm clock, and microwave oven.

A skilled abacus user can work out sums almost as fast as a skilled electronic calculator user.

HIGHER TECH

The story of gadgets follows the march of technology. The calculator began thousands of years ago as rows of pebbles, then wooden beads on wires, followed by hand-cranked mechanical versions, and then an electric motor design. From the 1940s, electronics and microchips not only shrank the calculator to pocket size but also started a whole new area for "gizmology." It has led to the ultimate gadget of our times—the computer.

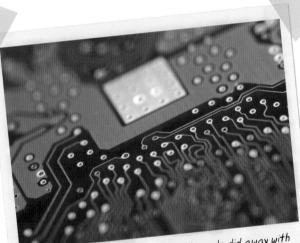

Mass-produced printed circuit boards did away with the labor-intensive task of installing wires by hand

Personal music players are small enough to be carried wherever we go.

SMALL IS BIG

The market for new gadgets seems endless. Advertisements continually announce smaller size, less weight, better batteries, easier controls, and more add-ons and features. Sometimes existing parts are miniaturized for a new use, like the micro versions of computer hard drives developed for personal music players. In other cases, a different area of technology is devised, such as flash digital memory sticks or pens for music players.

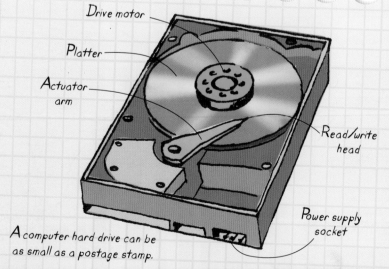

Drive motor

Platter

Actuator arm

Read/write head

Power supply socket

A computer hard drive can be as small as a postage stamp.

The gadgets featured in this book are Internet linked.
Visit www.factsforprojects.com to find out more.

Wide open spaces are recorded by a camcorder microchip smaller than your fingernail.

SIZE MATTERS

Some technologies resist the shrinking trend. Optical or light-based gadgets such as cameras and camcorders have to be a certain size, otherwise they could not receive enough light rays to record and capture the scene.

Manipulating light rays makes things look nearer or farther

Concave lens to front

Camera zoom lens at telephoto setting

Concave lens to rear

Camera zoom lens at wide-angle setting

NO LIMITS?

Will there come a time when all possible gadgets have been invented? That's very doubtful. Few people 25 years ago predicted the success of sports-playing computers that keep you active, brain-training video consoles, car satnavs, and Internet-enabled, do-everything mobile phones. What will the next 25 years bring? Perhaps a new generation of plug-in brain chips, a TV screen fitted inside the eyeball, and thought-wave communicators.

We can be sure of one thing—people will always want the most cutting-edge gadgets, so the well-worn phrase, "I want one of those" will always survive!

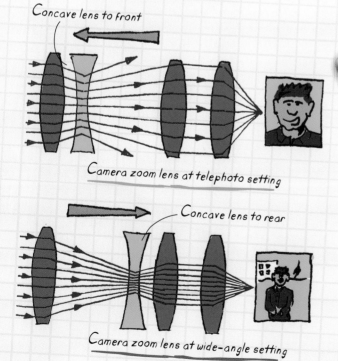

There are even gadgets to cause brain strain, such as this Nintendo DS gaming device.

CALCULATOR

Not long ago, people performed calculations with pencil and paper, a machine as big as a table with many gears, or in their heads. In the 1960s, handheld calculators were among the first small electronic gadgets. Today, calculators are used in offices, schools, factories, and homes across the globe.

Eureka!

A calculating machine called the abacus was used in ancient Greece and Rome over 2,000 years ago. It had rows of pebbles or beads in grooves on a table or in sand.

What next?

Scientists are testing foldable calculators with the keys and display built into thin, flexible cloth like a small handkerchief.

Microchips are also called ICs (integrated circuits), because they contain microscopic components, such as transistors and resistors, already integrated on one wafer-like chip of silicon.

Cover The outer cover is usually made of hard plastic to resist scratches and knocks.

✳ PORTABLE POWER

Early calculators had LED (light-emitting diode) displays that needed plenty of battery power. With the introduction of LCD screens, which use much less electricity, enough power for a small calculator could be provided by a photovoltaic cell, or solar cell, which turns light energy into electricity. Most modern gadgets have built-in secondary electrical cells—rechargeable batteries.

PCB (printed circuit board) Electronic components are linked by metal strips imprinted onto a board made of a green plastic-like substance that is an electrical insulator (does not carry electricity).

Microprocessor The brain of any electronic device is a chip called the microprocessor, or CPU, central processing unit. It carries out calculations from instructions fed into it via the keys.

Batteries range in size from smaller than a button to larger than a suitcase

The most complex microprocessors or CPUs have more than ten million components on a chip as small as this "o."

To find out more about LCD and how it works, visit
www.factsforprojects.com and click the web link.

Specialist calculators are used in
many areas of science, engineering
and even sports. They can work
out the amount of materials needed
to build a skyscraper, predict the
weather, or tell you how much air you
have left when scuba diving.

Display window

LCD The liquid crystal display can be
backlit for dim conditions, rather than using
reflected daylight. However the backlight
uses more electricity than the display.

Rubber pads

Keys The usual
numerical keys have
standard positions.
There is also a row of
arithmetical function
keys, usually on the
right, to add, subtract,
divide, and multiply.

Ribbon connector

Battery compartment Long
life or rechargeable batteries are
needed for bigger calculators used
for long periods, especially in dim
conditions where a solar cell could
not supply enough electricity.

Back panel

✳ How do LCDs work?

A liquid crystal can polarize light rays,
twisting them so they undulate (go up and
down) at the same angle, depending on
whether electricity flows through them.
Switched off, daylight enters the LCD at
the front, goes through all the layers,
bounces off the rear mirror, and comes
back out. Switched on, the liquid crystal
polarizes all the light rays into one certain
angle. Two other polarizing layers at the
opposite angle then block these rays.
No light can come back out, and the
display looks dark.

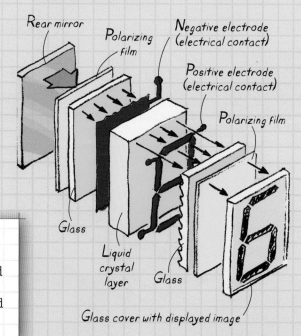

Rear mirror

Polarizing film

Negative electrode (electrical contact)

Positive electrode (electrical contact)

Polarizing film

Glass

Liquid crystal layer

Glass

Glass cover with displayed image

FLAT SCREEN

Flat screens are in computer monitors, TVs and closed-circuit TVs, displays on digital cameras, camcorders, satnav units, phones, and many other devices. There are two main types of flat screen technology—LCD (see page 9) and plasma (see below). Flat screens took over in the 1990s from older, heavier, boxlike, glass-screen displays known as CRTs (cathode ray tubes). A CRT uses far more electricity than a flat screen.

Eureka!

The first purely electronic television systems, with no moving parts, were developed in 1920–1930 by Hungarian engineer Kalman Tihanyi and Russian-American inventor Vladimir Zworykin.

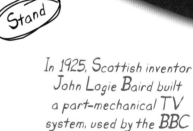

Clear cover

Screen format Most flat screens have a 16:9 aspect ratio for widescreen viewing, where the screen height is 9/16 of its width.

✳ How do PLASMA SCREENS work?

A plasma screen has millions of tiny compartments or cells and two sets or grids of wirelike electrodes at right angles to each other. Each cell can be "addressed" by sending electric pulses along two particular electrodes that cross at the cell. The electric pulse heats the cell's gas into a form called plasma, which makes an area of colored substance, the phosphor, glow for a split second. Millions of pulses every second at different "addresses" all over the screen create the overall picture.

Remote sensor A small infrared sensor detects the invisible IR (heat) beam from the remote-control handset.

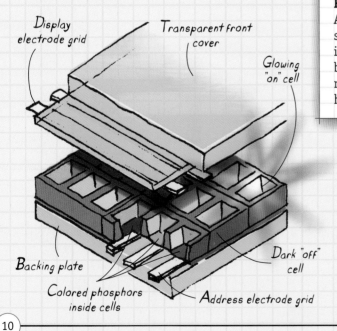

Display electrode grid

Transparent front cover

Glowing "on" cell

Backing plate

Dark "off" cell

Colored phosphors inside cells

Address electrode grid

Stand

The largest LCD flat screens measure 108 in (274 cm). That's the distance from corner to corner diagonally across the screen, which is the traditional way of sizing screens.

In 1925, Scottish inventor John Logie Baird built a part-mechanical TV system, used by the BBC from 1929 to 1937.

What next?

HD (high definition) screens have five to six times more tiny colored dots (pixels) than a standard flat screen. This gives a sharper, clearer, more colorful image and smoother, less jerky movements.

Read all about high-definition TV by visiting www.factsforprojects.com and clicking the web link.

The QuadHD flat screen has four to five times as many tiny colored dots (pixels) than an HD flat screen—but whether most people have good enough eyesight to tell the difference is unlikely.

Receiver A TV's receiver unit is part of the electronic circuit board. It tunes into different channels and then filters and strengthens the signals received by the aerial, ready for electronic processing. Computer monitors receive their signals already processed and so lack a receiver.

Ventilation slots The outer casing needs slots at the rear to allow heat to escape; otherwise, the thermal cutout makes the screen go blank to prevent components from overheating.

Electronics A typical HD flat screen TV has more than 1,000 microchips and other components.

Most LCD flat screens use TFTs (thin-film transistors). The transistor components for producing colored dots in the pixels are made within the thickness of the transparent screen.

Frame and backing The outer frame keeps the screen rigid and secure. It wraps around the electronics at the back. Some flat screens are less than 1 inch (2.5 cm) deep.

✳ DIGITAL BROADCASTING

Older television broadcasting was called analog. The information for the pictures and sounds was carried by the varying strength of the radio wave signals. In digital broadcasting, the information is coded in the form of on-off signals, millions every second. Radio waves used for one analog channel can carry up to ten digital channels.

Dishes receive digital signals direct from satellites in space

PERSONAL COMPUTER

Since the 1970s, personal computers have progressed from machines as big as a refrigerator, which only the wealthy or businesses could afford, to small personal computers (PC) found on and under desks in almost every home. A stand-alone PC works without being connected to a network with shared devices such as servers and other computers.

Eureka!

The first successful personal computer was the Commodore PET (Personal Electronic Transactor) in 1972. IBM's home computer, the IBM PC, was introduced in 1981 and set the standard for the computing industry.

What next?

Moore's Law says that computer power—the number of components on a microchip and how fast they work—doubles about every two years. Devised by Gordon Moore in 1965, it still holds true today.

Many modern computers have wireless peripherals (connected devices) that use short-range radio waves, such as Bluetooth technology, to carry information. This gets rid of trailing wires and allows them to be moved easily.

CPU The Central Processing Unit, or computer's brain, is connected to several other microchips. It carries out the main processing—altering data (information) according to the instructions from the application (program).

Cooling fan

RAM (Random Access Memory) chips

Card Smaller PCBs known as cards deal with signals going in and out, for example, to the display screen from the graphics card and to the loudspeakers from the sound card.

Hard drive The main memory disk that stores information permanently as micro-spots of magnetism is known as the hard disk or hard drive (see page 26).

✳ How do KEYBOARDS work?

All kinds of press button or key-activated gadgets, from cell phones to super computers, rely on a simple piece of technology similar to a light switch. Two pieces of metal conductor separated by a small gap are under each key. Pressing the key pushes them together so electricity flows. Each key has its own code of electrical pulses. A flexible membrane cover keeps out dust.

Top flexible membrane layer

Conductive strips not touching, open circuit

Finger presses down on keypad

Middle holes layer

Conductive strips make contact and complete circuit

Motherboard The main PCB (printed circuit board), which carries most of the major microchips and the connectors for other, smaller PCBs, is often called the motherboard.

For information on computing history and a game to play visit www.factsforprojects.com and click the web link.

Optical drive trays CDs (compact discs) and DVDs (digital versatile discs) are inserted here where they work using light. A laser beam reads tiny pits in their shiny surfaces (see page 31).

A typical home computer costs one-quarter of its price 20 years ago—and it is ten times more powerful.

Flat screen monitor

Untitled - Notepad
File   Edit   Format   View   Help

Untitled - Notepad

✳ The INTERNET

The global system of computers linked into an international network is known as the Internet. It began as a limited network for the U.S. military in 1969, was widened to research centers and universities in the 1970s, taken up by businesses in the 1980s, and opened for public access in the 1990s.

Pages and documents viewed on the Internet form the World Wide Web

IBM's Roadrunner, located at Los Alamos National Laboratory in New Mexico, is currently the world's fastest super computer.

Keyboard This is the main input device. It allows information to enter the computer as codes for alphanumerics—letters, numbers, and symbols such as + and &.

Mouse A laser beam or a rolling ball on the underside tracks the movements of the mouse. This makes the cursor or insertion point on the screen move in a similar way. The wireless mouse works by a Bluetooth short-wave radio link and so has no wire "tail."

ome computing did not really catch on until early mes machines appeared such as the Atari 2600 (1977) and Sinclair ZX Spectrum (1982). Games could be played in full color and with sound— amazing for the time!

DIGITAL CAMERA

A camera makes a permanent visual record of a scene, person, or object. Most cameras have a lens to focus the light rays for a clear image. A digital still camera (recording a still, split-second moment in time) turns the pattern of light rays into on/off electronic signals stored in a microchip.

Eureka!

Early digital cameras included the Fuji DS-1P in 1988 and the Dycam M1 in 1990. Kodak introduced its first digital camera in 1991 with the DCS-100.

What next?

Early digital cameras recorded images of about one megapixel. For a typical camera in the 2000s, this rose to 6 megapixels, then 8. It will continue to increase as CCD chips improve.

Digital cameras may not capture detail or color variation as a film camera can. However, they display the stored image, which can be deleted if needed. They also hold more images than a roll of film.

Autofocus An invisible infrared beam bounces off the object in front of the camera, which detects the time taken for it to return. This shows the object's distance for lens focusing.

Aperture This hole is made larger or smaller to let in more or less light, depending on brightness conditions.

✳ MEGAPIXELS and RESOLUTION

A megapixel is one million pixels, which are picture elements—tiny areas or spots in the whole picture. Each pixel is made up of red, green, and blue light. In various combinations, these produce all other colors. Mixed together, they make white. Packing more pixels into a given area makes the image clearer and more detailed, which is known as going from low resolution to high resolution.

Lens The lens system has several curved pieces of glass or plastic that move forward or backward to focus the image, depending on the object's distance.

"Take" button

FinePix

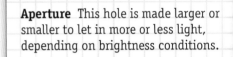

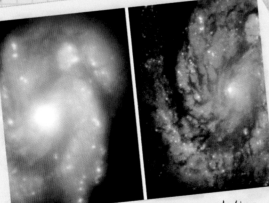

A space galaxy at low resolution (left) and higher resolution (right)

Battery

Shutter Pressing the "take" button opens a door-like flap for a split second to allow light through to the CCD (see page 18).

Many digital cameras can take simple sequences of moving images or videos, and record sound, too.

Memory card

For useful hints and tips on how to use your digital camera, go to www.factsforprojects.com and click the web link.

A traditional camera stores images as patterns of chemical changes, containing silver, on a flexible cellulose-based roll of photographic film.

Viewfinder A small LCD display shows the view that the lens sees, which is the image that will be stored.

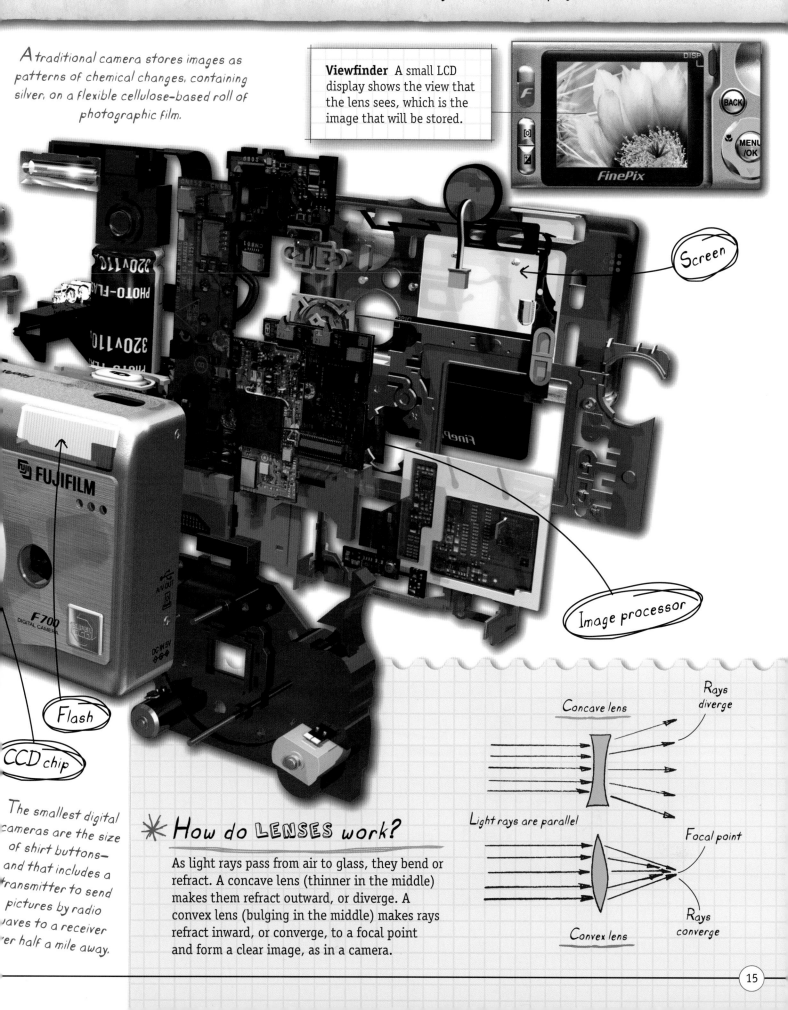

Screen

Image processor

FUJIFILM

F700
DIGITAL CAMERA

Flash

CCD chip

The smallest digital cameras are the size of shirt buttons— and that includes a transmitter to send pictures by radio waves to a receiver over half a mile away.

✳ How do LENSES work?

As light rays pass from air to glass, they bend or refract. A concave lens (thinner in the middle) makes them refract outward, or diverge. A convex lens (bulging in the middle) makes rays refract inward, or converge, to a focal point and form a clear image, as in a camera.

Concave lens

Rays diverge

Light rays are parallel

Focal point

Rays converge

Convex lens

CAMCORDER

The camcorder is a camera-recorder specialized to take video—very fast split-second sequences of still images that look like moving pictures—and record them for later playback. It also records sounds, which are linked in time, or synched (synchronized), to the images. Some camcorders record on magnetic tape, either analog or digitally. Others are digital, and use very small hard drives or electronic microchips in memory cards.

Eureka!

The first video cameras were cinema cameras using photographic film. Made in the 1880s, they recorded early movies or cinema motion pictures. Small hand-held versions became popular from the 1950s.

What next?

One day, a light-detecting electronic chip that works as a simple video camera could be put into the eye, so you can make a permanent record of everything you see.

Zoom gears A small electric motor and gear system move the lenses so that the camera can focus and zoom in to show a small area larger (see right).

Lens cover

✳ SATELLITE VIDEOPHONE

Unlike networks for local and cell phones, satellite video phones have a radio link to a satellite orbiting in space. The transmitter-receiver is housed in a laptop-sized case. A headset carries the camera to record the view, the microphone for sound, and a display screen and headphones, either for what is being recorded, or to relay video and sound sent through the satellite link to the wearer.

Casing The scratchproof outer casing provides some protection against bumps.

Main processors The main microchips process the digital signals from the CCD into a form that is easily stored in the memory device.

The headset camera (left) records what is in front of the person's face

The highest-quality digital formats include MiniDV and Digital Betacam. They lose much less quality when the video recording is later copied and edited, compared to an analogue recording.

Create your own digital movie by visiting www.factsforprojects.com and clicking the web link.

Microphone A small microphone (see page 22) picks up sounds during image recording. On some cameras, it can be detached for accurate aiming, linked to the camcorder by a wire or radio waves.

Small and lightweight, early camcorders were developed for "on-the-spot" recording and reporting for TV.

Viewfinder The LCD screen allows you to see what has been recorded and delete it, or edit it by picking out only the parts you want to save.

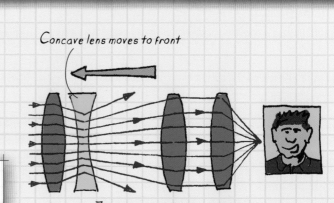

Concave lens moves to front

Zoom lens at telephoto setting

Concave lens moves to rear

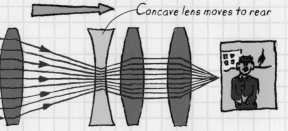

Zoom lens at wide-angle setting

✳ How does ZOOM work?

Zooming in, or telephoto, enlarges a portion of the scene, while showing a relatively smaller area of the whole scene. Zooming out to wide-angle takes in more of the scene but reduces its magnification so things look smaller. The zoom system uses a movable concave lens (one that is thinner in the center than around the edges) to spread out or diverge the light rays (see page 15). Some zooms are worked by an electric motor, others are twisted or pulled by hand.

Battery

LCD display

USB memory stick

Screen controls The screen has brightness, contrast, color balance, and other controls. You can adjust these to view it comfortably out in the glaring sunshine or indoors in a darkened room.

Most photographic film video cameras take 24 separate still pictures or frames every second. Camcorders usually take 30 frames per second. When the recording is played back, your eyes cannot see the individual frames separately. Your brain blurs them together to give the impression of smooth movement.

Webcams are small, simple, video cameras that do not record their images but feed them straight into a computer or digital network.

SCANNER

The key to digital technology is to turn anything—pictures, videos, sounds, speech, written words, and numbers—into coded sequences of on/off electrical pulses. We write these as the digits (numbers) 1 and 0, and they are the language of computers. An image scanner digitizes a picture by detecting the color and brightness of every tiny spot, one spot after another, and converting this information to digital code.

Eureka!

The first image scanners appeared in the 1960s as part of the work to develop body scanners for medical use. Scanners were used by businesses in the 1980s and became available for home use in the early 1990s.

What next?

Three-dimensional (3D) scanners consist of two digital cameras mounted on a frame. They are moved over and around an object to make a 3-D view in the computer memory.

Scan technology involves looking at and detecting a row of tiny areas in a straight line, then moving along slightly and doing the same for the next line, and so on—usually thousands of times.

Light source A very bright light, as pure white as possible, shines onto the image. Its light rays reflect off the image with the color and brightness of every part of that image.

Drive motor housing

PCB

Rail The scan head slides back and forth along a metal guide rail. It is driven by a toothed belt that meshes to a sprocket (gear wheel) for very accurate movement.

Drive gear

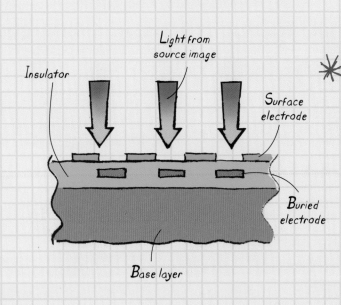

Insulator

Light from source image

Surface electrode

Buried electrode

Base layer

✴ How does a CCD CHIP work?

A CCD (charge-coupled device) is a microchip with millions of tiny wirelike electrodes forming a crisscross grid. When a ray of light hits the area or boundary between two neighboring electrodes, it causes a particle called an electron (an outer part of an atom) to jump between them. This causes a tiny pulse of electric current at that point on the chip—light has been converted into electricity.

Control buttons The scanner has some manual controls, such as resetting the scan head, to scan only a small area, or to make a direct copy. Most of the controls are in the computer application to which it is linked.

For some great tips on scanning at home, visit www.factsforprojects.com and click the web link.

The detail that a scanner picks up is measured as dpi, (dots per inch). This shows how many dots of color are detected in a line one inch long.

Tray

Direct output (copy)

If large areas of an image are the same color, the computer can save memory by giving them all a shortened version of the code for that color, rather than the full code for every tiny spot of the area. This is one way of compressing a computer file to make it smaller.

Sealed case The scanner's internal parts are sealed to keep them dustproof, since buildup of dust on the mirrors, lenses, and CCD would greatly affect the quality of the scan.

Power button

A red shirt is changed to blue— or the other way around?

Flatbed The image is laid flat on a glass sheet for flatbed scanning. Some scanners have a different design where the image is wrapped around a drum that revolves past the scanning head. Drum scanners can produce a far more detailed scan.

Main housing Inside the scanner's main case is a mirror that reflects light that has already been reflected from the image. The mirror reflects the light toward the CCD, which turns the patterns of color and brightness in the light rays into a corresponding pattern of electronic digital signals.

✳ VSFX

Using keyboard instructions, computers can change or process words. The same can be done with pictures, known as image processing or visual special effects (VSFX). For example, the computer can be instructed to change all areas of a certain shade of red in an image to a blue color.

PRINTER

A typical home printer makes hard copy of what is on the computer screen, on a sheet of paper or card where it is permanent or hard. In some ways, printer technology is the opposite of the image scanner (see page 18). The printer builds up the image as many tiny dots of ink to form a long row or line, and then another line next to it, and so on.

Eureka!

Computer printers for home use were developed in the 1960s. The print head had a gridlike set or matrix of tiny pins that pressed an inked ribbon onto paper, similar to typewriter technology. This was called the dot-matrix printer. It could only produce marks the color of the ribbon.

What next?

Skin printers are small rollers that squirt out ink as they pass over the skin. They make images of any kind, usually from washable inks—like having an almost instant temporary tattoo.

There are two main types of design for inkjet print heads. In the fixed-head version, the nozzles are permanent and need cleaning occasionally. In the disposable head type, a new set of nozzles comes with each cartridge of ink or toner.

Rollers Driven by an electric motor, the roller system moves the paper past the print head a strip at a time and the print head produce each strip of the image.

Cover

Output

✳ 3-D magic

Instead of the usual two-dimensional or 2-D printing on a flat surface, 3-D printing builds up solid objects layer by layer, with depth or height as the third dimension. A powder spreads out in the printer chamber, then a laser warms it only in the area of the print, so its particles stick together. The next layer is added on top, and this layer fuses to the one below. When the layering is complete, the loose powder is removed. This method is used for prefabricating—making models and prototypes of complex shapes.

A laser beam heats the powder in a 3-D printer chamber.

Output tray Printed sheets stack up here. It is wise to let them dry separately, uncovered, where air can get to the ink. Otherwise, they may smudge as they slide and press together.

Read more information on how printers work by visiting www.factsforprojects.com and clicking the web link.

Glossy photo paper has a shiny finish that gives printed images a vivid look with brighter, more intense colors and a sharper, clearer quality.

Ink cartridges The colored inks or toners (powder) usually come in small sealed containers. These fit into a slot and are then punched open to release the ink, and pressurized to get the ink into the nozzle area.

Paper tray New sheets of paper are fed one at a time through a tiny gap by rubber rollers.

Case

Ink and paper must be matched so the paper has enough absorbency to soak up the ink, but the ink is not too runny to spread and smear through the paper.

Drive rollers These are spun by a stepper motor—an electric motor that moves the paper along in tiny amounts or increments. The amount of movement depends on the resolution—the closeness of the printed dots.

Ribbon connector

Power button

Print head This zooms back and forth along a guide bar or rail, squirting out tiny jets of ink as it makes one line. Then the paper moves slightly, ready for printing the next line.

Some inkjet cartridges can be recycled by being cleaned and refilled with ink (liquid) or toner (particles).

✳ How do PRINT HEADS work?

The print head moves across the paper, squirting out lots of tiny ink drops as it goes. To produce each drop, a tiny heating wire makes the ink expand into a bubble, forcing a small amount through the narrow nozzle at the tip. Any color can be made by different combinations of the three primary ink colors—cyan, magenta, and yellow. All three of these colors make black. However, ready-made black ink saves using up the colored inks.

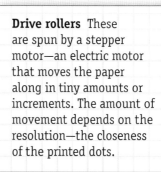

Close-up of print head
Heating element — Tube containing ink
Nozzle inside print head
1. Heating element heats up forming a bubble in the ink
2. Growing bubble forces ink from nozzle onto paper — More ink flows from cartridge
Magenta, yellow, black, and cyan (right) inks are used.

RADIO MICROPHONE

A microphone changes patterns of sound waves into similar patterns of tiny electrical signals, which are fed into an amplifier to make them stronger. The long wire trailing from an ordinary microphone to the amplifier means that performers cannot move about with freedom—and they might trip over it. The radio microphone sends its signals to the amplifier by radio waves.

Eureka!

The first microphones were in the first telephones, made by Alexander Graham Bell and his co-workers in the 1870s. By the 1920s bigger, better microphones were in regular use for radio broadcasts.

What next?

A small pellet-like microphone can be placed in a false tooth to transmit what you say by radio—but you have to turn it down when you eat!

Performers often check a microphone by saying, "testing, one, two." These words contain common voice sounds, such as "s" and a hard "t."

Body or barrel

Batteries The barrel or main body of the microphone is shaped for being gripped in the hand and housing the batteries. A warning light shines when battery power begins to fade. To save the batteries, the mic can be switched to mute, which means it goes silent but is ready for use again in an instant.

Transmitter and antenna The radio transmitter sends out coded radio waves from its antenna (aerial), carrying the sound information to a receiver that is usually less than 300 feet (91.4 m) away.

SIZE AA BATTERY

SIZE AA BATTERY

DURACELL
MN 1500 LR6 1.5 VOL
CAUTION: DO NOT CONNECT IMPROPE
CHARGE OR DISPOSE OF IN FIRE. BAT
MAY EXPLODE OR LEAK.
Made In

Hand grip

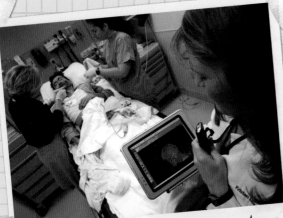

In a hospital, a Bluetooth microphone links by radio to its base next door

✳ What is BLUETOOTH?

To avoid too many tangled and trailing wires, electrical devices can send information between themselves using radio waves, known as wireless. Bluetooth is not the name of a particular make or type of device. It is the label for the way the information is coded into radio signals and a standard for the speed, quality, strength, and reliability of the short-range wireless system. All Bluetooth devices can communicate to each other in the standard way, rather than different manufacturers using their own systems.

"Bugs" are small listening devices made up of a tiny microphone and radio transmitter. They are often used for spying and secret investigations.

Find out more information about different types of microphone
by visiting www.factsforprojects.com and clicking the web link.

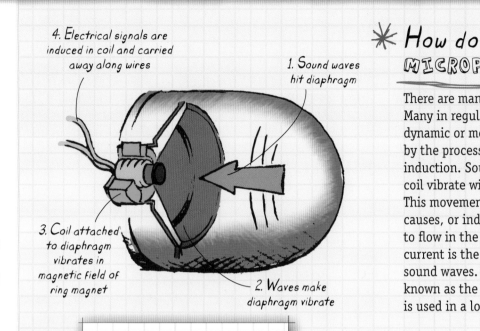

4. Electrical signals are
induced in coil and carried
away along wires

1. Sound waves
hit diaphragm

3. Coil attached
to diaphragm
vibrates in
magnetic field of
ring magnet

2. Waves make
diaphragm vibrate

As well as
moving-coil
microphones, there
are many other
kinds including
condenser, laser,
electrostatic,
carbon-granule,
and piezoelectric
mics. Each has
its own benefits,
such as lightweight,
durability, or
high-quality sound
reproduction.

✳ How do MICROPHONES work?

There are many kinds of microphones.
Many in regular stage use are called
dynamic or moving-coil mics and work
by the process of electromagnetic
induction. Sound waves make a wire
coil vibrate within a magnetic field.
This movement of a wire near a magnet
causes, or induces, electrical currents
to flow in the wire. The pattern of the
current is the same as the pattern of
sound waves. The opposite process,
known as the electromagnetic effect,
is used in a loudspeaker (see page 24).

Magnetic coil The coil attached
to the diaphragm is wound
from many turns of ultrathin
wire. The more turns it has, the
stronger the electrical signals
it generates, and so the clearer
the sounds are.

Permanent
magnet

A unidirectional
microphone
picks up sounds
from only one
direction.

Muffler or guard A foam cover
over the diaphragm prevents
the wind and air currents from
rattling the diaphragm. But
at the same it is acoustically
transparent, allowing through
sound waves from the air around.

SHURE

Head

PCBs Printed circuit
boards filter, convert,
and clean up the
electrical signals
from the wire coil,
ready to feed to
the transmitter.

Diaphragm The thin, flexible diaphragm shakes
very fast, or vibrates, when sound waves bounce
off it. This makes the wire coil attached to it
vibrate with a similar pattern of movements.

An omnidirectional
mic detects sounds
from all directions.

LOUDSPEAKERS

A loudspeaker or speaker turns electrical signals into corresponding patterns of sound waves—the opposite of a microphone (see page 22). To make the electrical signals strong enough to power a loudspeaker, they usually have to be boosted or increased by an amplifier or amp.

Eureka!

Like early microphones, the first loudspeakers were developed by Alexander Graham Bell and his colleagues for the first telephones during the 1870s.

What next?

Tiny Bluetooth-type radio earphones could be implanted into the human ear and activated by brainwaves. It may then be possible to switch on a radio link and listen just by thinking.

The massive speakers at big events need thousands of watts of power. The tiny speakers in earphones need just a few thousandths of a watt.

Unlike many modern electronic devices, loudspeakers are analog. They use electrical signals of varying strength rather than on/off digital signals.

Coil The wire coil inside the magnet, also known as the voice coil, has many turns of very thin wire. This makes it move as much as possible even with very weak electrical signals.

Ring magnet

1. Electrical signals are sent to the speaker

3. Diaphragm connected to coil enlarges the vibrations, causing sound

2. Magnet and coil attached to diaphragm convert electrical signals into vibrations

4. Sound travels from speaker out into air

✳ How do MOVING-COIL LOUDSPEAKERS work?

A loudspeaker uses the opposite principle to electromagnetic induction, which is known as the electromagnetic effect. When an electric current passes through a wire coil, this becomes an electromagnet with its own magnetic field. In a speaker, the field interacts with the constant field of a ring-shaped permanent magnet around it, alternately pulling (attraction) and pushing (repulsion). This makes the coil vibrate, which passes its vibrations to the diaphragm, that then produces the sound waves.

Stand

Diaphragm Also called the cone, it is usually made of stiff plastic or specially treated paper. It has a flexible outer rim so it can vibrate freely within its frame.

Carry out a simple activity and learn more about making and hearing sounds by visiting www.factsforprojects.com and clicking the web link.

* Tiny SPEAKERS

Earphones are tiny loudspeakers that fit in the ear. The magnets in earphones are made of special alloys, combinations of metals, that have a very strong magnetic field but are also light and not damaged by bumps and vibrations. These alloys include the rare metals samarium and cobalt, which make earphones with powerful low frequencies for a deep bass sound. Headphones have slightly larger speakers and padded cups that fit over the ears to keep out unwanted sounds.

The term loudspeaker can be used for the box, cabinet, or container housing one or more actual speaker units, which are then known as drivers.

Earphones let us listen without disturbing others too much

Tweeter This smaller loudspeaker produces shrill notes or high frequencies such as cymbals and the "s" sounds of the voice.

Subwoofer loudspeakers (or subs) produce sound waves that are too deep or low for the human ear to detect. However, we can feel them in our bodies as a deep thud or thump.

Grill A protective cover prevents damage to the loudspeaker or driver unit but lets sound waves pass through.

iPod

Dock

Woofer The larger loudspeaker produces deep notes or low frequencies produced by drums and bass guitars.

Apart from moving-coil or dynamic loudspeakers, there are also electrostatic, ribbon, and piezoelectric (crystal-based) types for special applications.

Some very high-quality loudspeaker systems have five driver units for different frequencies or pitches of sound: subwoofer, woofer, midrange, tweeter, and ultratweeter.

PERSONAL MUSIC PLAYER

Various manufacturers make portable personal music players, also called digital audio players. Those produced by Apple are called iPods, while many others are known by the general name of MP3 players. The term MP3 refers to the way the sounds are changed into digital code and compressed to take up less memory space without loss of quality.

Eureka!

The iPod took little more than one year to develop under the watchful eye of Apple chief Steve Jobs. It was launched in 2001 with the catchphrase "1000 songs in your pocket" and has been a best seller ever since.

What next?

MP4 is faster to transfer than MP3, works with audio and video, uses less memory, and is more suitable for streaming (transferring "live" in real time) over the Internet.

MP3 stands for mpeg audio layer 3. The term mpeg (Moving Pictures Expert Group) refers to a team of electronics experts who determined how images and sounds could be changed into digital electronic form using a series of mathematical sums and formulas known as algorithms.

Battery pack

Display The LCD screen shows menus to choose which song to play, earphone volume, and many other features.

Slim case

Hard drive This mini-version of a computer hard disk stores all the audio information, which is usually measured in gigabytes (GB) of memory space.

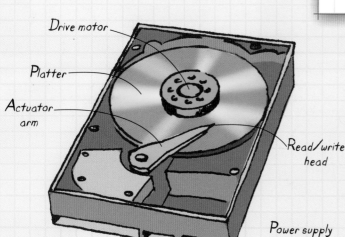

Drive motor
Platter
Actuator arm
Read/write head
Power supply socket

✳ How does a HARD DRIVE work?

A hard drive (hard disk) is one or a stack of spinning discs or platters, each coated with a very thin layer of magnetic substance. Read-write heads on arms swivel over the disk's surface, hardly touching. They write information in specific places by creating patterns of magnetic spots, or read information by detecting these spots.

Shockproofing The hard drive in a digital player is tough. But it is also covered by rubber strips or similar packing to cushion it from bumps and jolts.

Discover everything you need to know about MP3 players by visiting www.factsforprojects.com and clicking the web link.

EXPLODED VIEW OF APPLE IPOD

Some digital music players with display screens have video games such as Brick, Solitaire, and Music Quiz.

Flash memory is non-volatile, which means it doesn't need an electricity supply to refresh or maintain it. It should stay in a microchip for years with no electrical power.

Album artwork

Motherboard

MENU

✳ FLASHY MEMORY

Some MP3 players use flash memory, which is a microchip that holds information that can be deleted and reprogrammed. Flash memory chips or sticks are smaller and cheaper than a tiny hard drive or microdrive but hold less information. However, they also use less electricity and allow batteries to last longer than with a microdrive. Most flash chips are protected in plastic cases and plug into the USB (universal serial bus) socket of a player, computer, or network.

The USB memory stick plugs into many electronic devices

Click wheel Instead of pressing buttons, a user presses or strokes parts of the wheel to play and pause, skip forward or back, adjust the volume, choose a menu item, and return to the previous menu.

Touch sensors A series of flat components called capacitors detects when an object, such as a finger, is near and which way it moves. The presence of the object alters the amount of electricity that the capacitor can hold.

The forerunner of the iPod/MP3 player was the portable laser-based CD player. It was more than ten times bigger and heavier than an MP3 player. Moving it often made the compact disc skip.

27

VIDEO GAMES CONSOLE

Computer games consoles and video games machines have been around since the 1960s. Early versions had just a few games, like tennis with two bats and a ball—in black and white! The latest consoles have hundreds of games in stunning fast-action 3-D color. You can play yourself, against friends, or online with someone on the other side of the world.

Eureka!

The first computer game using graphics (pictures and symbols rather than words) was noughts and crosses in 1952. *Tennis* followed in 1958, and the first true computer game *SpaceWar!* arrived in 1962.

What next?

For health reasons and to prevent people from becoming obsessed with video games, some people believe the games should have built-in breaks for ten minutes every hour.

SONY PLAYSTATION

Space Invaders was one of the most popular games ever. It was launched in 1978 in video gaming arcades and then for home machines. No other arcade game has been played as much.

Main board

✳ Virtual WORLDS

Video games on the Internet have become so detailed and complex that they are almost like leading an alternative existence in the virtual world of computing. Avatars (a name derived from the Hindu religion) are characters or representations that gamers use for themselves, varying from humanlike forms to animals, monsters, robots, machines, or simple symbols or icons.

Chips The many microchips include a clock chip that ensures all the circuits work together properly and signals are sent between them on time.

Video games consoles are now in the seventh generation with models such as the Sony PlayStation 3, Microsoft Xbox 360, and Nintendo Wii.

Gamers gather at big meetings to play the latest releases

Optical disc drive Advanced games machines can play DVDs and Blu-ray discs and be used as a disc player to watch movies on a high-definition (HD) television or computer display.

Find out everything you need to know about video games and consoles
by visiting www.factsforprojects.com and clicking the web link.

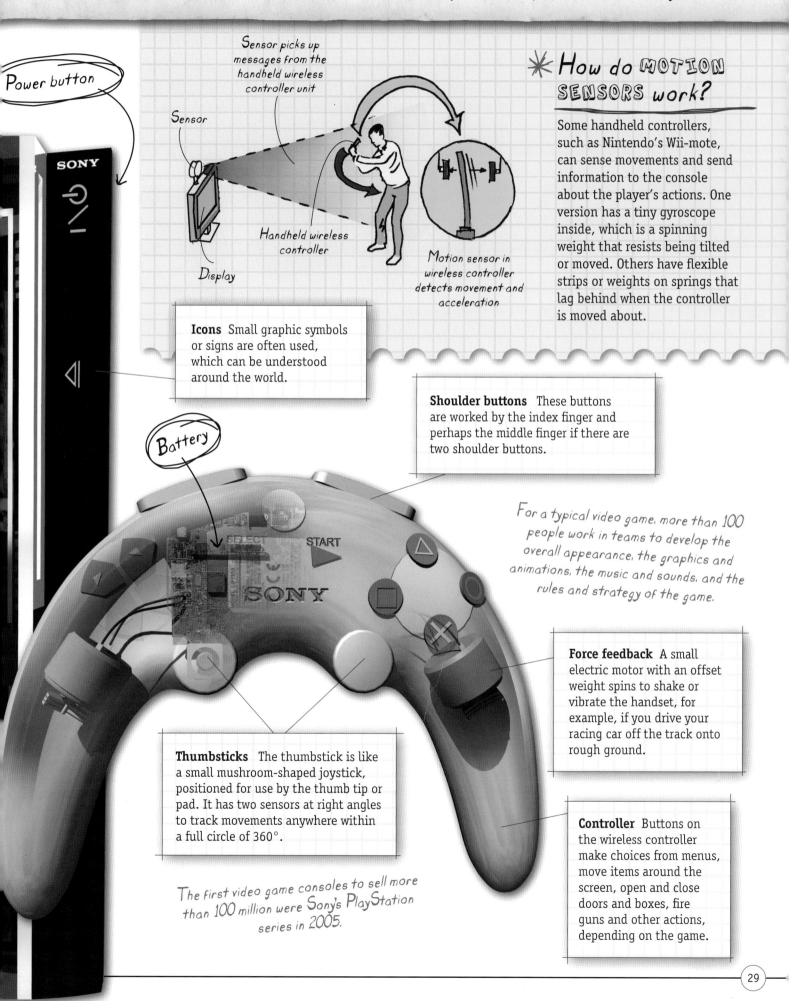

Power button

Sensor picks up
messages from the
handheld wireless
controller unit

Sensor

Handheld wireless
controller

Display

Motion sensor in
wireless controller
detects movement and
acceleration

SONY

Icons Small graphic symbols
or signs are often used,
which can be understood
around the world.

✳ How do MOTION SENSORS work?

Some handheld controllers,
such as Nintendo's Wii-mote,
can sense movements and send
information to the console
about the player's actions. One
version has a tiny gyroscope
inside, which is a spinning
weight that resists being tilted
or moved. Others have flexible
strips or weights on springs that
lag behind when the controller
is moved about.

Shoulder buttons These buttons
are worked by the index finger and
perhaps the middle finger if there are
two shoulder buttons.

Battery

SELECT   START

SONY

For a typical video game, more than 100
people work in teams to develop the
overall appearance, the graphics and
animations, the music and sounds, and the
rules and strategy of the game.

Force feedback A small
electric motor with an offset
weight spins to shake or
vibrate the handset, for
example, if you drive your
racing car off the track onto
rough ground.

Thumbsticks The thumbstick is like
a small mushroom-shaped joystick,
positioned for use by the thumb tip or
pad. It has two sensors at right angles
to track movements anywhere within
a full circle of 360°.

Controller Buttons on
the wireless controller
make choices from menus,
move items around the
screen, open and close
doors and boxes, fire
guns and other actions,
depending on the game.

The first video game consoles to sell more
than 100 million were Sony's PlayStation
series in 2005.

HOME THEATER

A trip to the theater to watch the new blockbuster movie on the big screen with loud sound all around is an exciting event. But in the past 20 years, home theaters have become bigger and better. The key parts are a widescreen display or television, a sound system with several loudspeakers, and the movie that is usually on a DVD player.

Eureka!

The first CDs were sold in the early 1980s. They were used for audio (sound), usually recorded music. Computer engineers then used CDs to store words, pictures, and software applications or programs, and data files. In recent years, CDs have been partly replaced for data storage by devices such as mini hard drives, microdrives, and memory cards, chips or sticks.

What next?

Scientists are in the process of developing holographic displays where the scenes and objects are projected as colored lights in three dimensions, including true depth rather than the illusion of depth on a flat screen.

Blu-ray discs look like bluish versions of a CD or DVD. They use blue laser light, which can read tinier spots than the usual red DVD laser. This enables the Blu-ray disc to hold more information (25 or even 50 GB), for movies in HD (high definition).

Speakers In a standard 5:1 loudspeaker setup, there is one subwoofer for very deep sounds that can be positioned almost anywhere. Left and right front speakers give a stereo effect. A central front speaker fills the gap for sounds coming from the middle. Left and right rear speakers provide a surround sound.

Many modern smash-hit movies, such as WALL-E, use CGI

✳ SUPER-REAL

Many spectacular films use CGI (computer generated images) for special effects amid real-life action or for an entire animated movie. Characters and scenes are built up as 3-D netlike "meshes" in computer memory, showing their basic shapes. The meshes can be manipulated and changed according to mathematical rules, such as when a character walks.

DVD player A standard DVD holds about 4.7 GB (gigabytes), which is up to six times more information than a CD can hold. That's enough for color pictures and sound for more than two hours of movie.

The first film screening was in Paris, France, in 1895. Brothers August and Louis Lumière showed ten 50-second films they had made, including one of workers leaving their factory.

For a complete guide to home theater systems, visit www.factsforprojects.com and click the web link.

Plasma flat screen

Wide screen The wide screen format of 16:9 (see page 10) has the correct proportions of width to height to fit in our field of vision—the complete area we can see with both eyes.

Component shelving

Connecting cables There are several types of connecting cables including elongated 21-pin SCART plugs, component video leads with red, green, and blue plugs, small round S-video plugs, DVI (Digital Visual Interface) plugs as used for computer monitors, and the latest 19-pin HDMI (High-Definition Multimedia Interface) design.

✳ How do CDs and DVDs work?

Compact discs and digital versatile (or video) discs are optical—they use light. Each disc has a spiral track of tiny hollows or pits with flat areas, called lands, between. The sequence of pits and lands contains the digitally coded information. As the disc spins, a laser beam reflects off the pits but not the lands. The flashes of reflection are picked up by a detector.

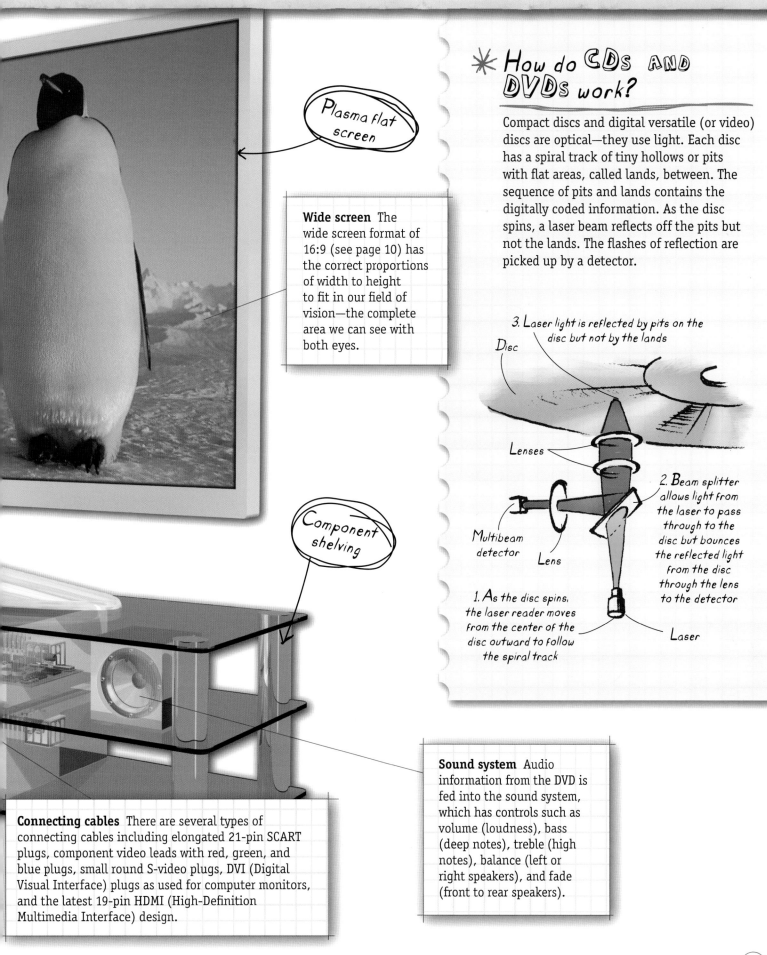

3. Laser light is reflected by pits on the disc but not by the lands

Disc

Lenses

Multibeam detector

Lens

2. Beam splitter allows light from the laser to pass through to the disc but bounces the reflected light from the disc through the lens to the detector

1. As the disc spins, the laser reader moves from the center of the disc outward to follow the spiral track

Laser

Sound system Audio information from the DVD is fed into the sound system, which has controls such as volume (loudness), bass (deep notes), treble (high notes), balance (left or right speakers), and fade (front to rear speakers).

VIRTUAL REALITY

Reality is real. Virtual reality is not—but it seems real to our senses such as vision, hearing, and touch. The best virtual reality (VR) systems feed information such as sights, sounds, and movements to the human brain and trick it into believing that things are real and actually exist, rather than being artificial. VR can be used for fun and entertainment, or for serious purposes such as training pilots to fly and surgeons to carry out medical operations.

Eureka!

Some of the first VR machines were flight simulators for pilots and air crew. In World War II (1939–1945) the Celestial Navigation Trainer, a massive machine at 45 feet (13.7 m) high, could hold an entire bomber aircraft crew training for night-time missions.

What next?

As VR equipment becomes faster and more complicated, more body senses can be stimulated. Tiny pellets of different odors can be released from the headset, for example, the smell of smoke when training firefighters to tackle a virtual inferno.

Screens Two screens give slightly different views of the scene, just like human eyes. The brain merges these into one 3-D view.

The Virtual Cocoon Room stimulates all the senses to take the person on a tour anywhere in the world—from the African grasslands to a deep cave.

Earphones Stereo sound is played through the earphones or headphones. Action that looks as if it is happening to the left of the wearer is accompanied by sounds that are louder in the left earpiece.

✳ VIRTUALLY FLYING

Full motion flight simulators show a widescreen view of what is outside, which changes according to which way the pilot guides the aircraft. The simulator (sim) also leans, tilts, and shakes using powerful, fast-acting hydraulic pistons under the seat, to recreate the movements of the aircraft. It can make a beginner feel very airsick!

Pilots train in a Boeing 727 sim

Virtual surgery can be extended into telesurgery or remote surgery. A surgeon in one place works a console that controls robot equipment operating on the patient in another place.

Play an online flight simulation game by visiting
www.factsforprojects.com and clicking the web link.

Headset The display screens and earphones or headphones are fixed to the VR headset. It should fit comfortably, so the wearer is not distracted by its pressure and gradually forgets about it.

Artificial reality was a term invented by American computer-based artist Myron Krueger, in the 1970s. Virtual reality goes back 50 years earlier to French writer, actor, and director Antonin Artaud.

Motion and pressure sensors In VR gloves, tiny sensors at different locations detect how much pressure the wearer is putting on an object. The VR computer calculates the strength of grip and makes the object move and react appropriately on the screens.

Wireless link The VR headset, gloves, and perhaps even a whole body suit are linked by radio to the main console, so that the wearer can move and react freely.

The first VR headsets were built in the late 1960s. They were so heavy that they had to be hung from a frame above to avoid crushing the wearer!

✳ How does STEREO VISION work?

Each of a person's eyes see a slightly different view of the world. This is stereoscopic vision. The closer an object, the more different these views. By comparing them, the brain works out the distance of the object. A VR headset's screens show two different views, one for each eye, with no area of overlap as when looking at a normal TV screen. The brain combines the separate views so that objects appear as if they are near or far.

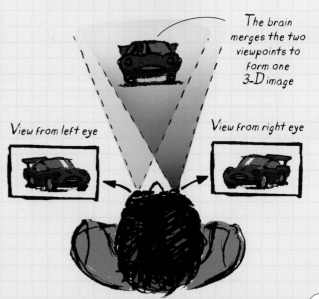

The brain merges the two viewpoints to form one 3-D image

View from left eye

View from right eye

CELL PHONE

It's hard to imagine life without the cell phone. Just 20 years ago, these handy gadgets were three times bigger and cost five times more—and text messaging was almost unknown. Today's models do not continue to shrink so much in size, as they did in the past. Rather, they pack in more and more functions, such as games, camera, video and sound recording, satnav, Bluetooth short-range wireless link, Internet access, music player, radio, and television.

Eureka!

In the 1980s, early cell phones were the size of bricks and just as heavy. Apart from progress in microchips and radio circuits, one of the main advances has been batteries that are smaller, lighter, and last longer.

What next?

Technically, it is possible to make a cell phone smaller than your little finger. However, the screen, icons, and buttons would be too small to see or manipulate. Advances in voice control may overcome this problem.

EXPLODED VIEW OF APPLE IPHONE

In 2008, about two cell phones in every five sold were made by Nokia.

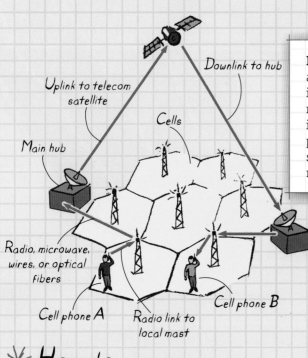

Uplink to telecom satellite

Downlink to hub

Cells

Main hub

Radio, microwave, wires, or optical fibers

Cell phone A

Radio link to local mast

Cell phone B

Icons An icon is a small picture, image, or symbol representing information or a particular function such as Send text message.

✷ How do CELL PHONES work?

Each area or cell has a radio transmitter-receiver mast that regularly sends out its identification code. A cell phone detects the strongest code and sends outgoing talk or text messages to it. The mast links by radio, microwaves, or wires to a main hub, which communicates with the entire telecom network including satellites. The network finds the receiver cell phone and sends the message to the nearest mast—which could be in the next cell!

In clear, flat, open areas, a typical cell phone has a range of about 25 miles (40 km) to the nearest transmitter-receiver mast.

NiMH cell The nickel-metal hydride cell is a small, but powerful, and long-lasting rechargeable battery. Larger versions power electric cars and trucks.

Discover how cell phones send and receive calls by visiting www.
factsforprojects.com and clicking the web link.

In areas of great natural beauty, cell phone transmitter-receiver masts may be disguised as trees, so as not to ruin the view.

Texting is convenient and cheaper than making calls

✳ WHY TEXT?

Text messaging, also known as SMS (Short Message Service) carries information of up to 160 characters (numbers, letters, and symbols such as & and @). A text takes only a split second to send and is cheaper than talking. Also, you do not hold the phone to your ear. Some people worry that the radio waves it transmits could damage the ear or even the brain. So far, medical tests show that there is no evidence of harm from using a cell phone.

Protective screen

Touch screen The touch screen does away with physical buttons on the phone interface by showing button icons to press. This is more adaptable, and can change the button icons according to the phone's mode, such as address book, gaming, or texting.

On/off/sleep The main power button puts the phone into energy-saving "sleep" mode when pressed. It turns the phone on or off when held down for a longer time.

PM

Photos

Camera

Weather

Maps

iPod

Settings

Safari

Notes

Calculator

Mail

Clock

Flash

iPod

Metal case

Phone icon

Connections Sockets link the phone physically to a computer or network, to download (send away) or upload (receive) information such as images, sounds, text messages, and games.

Cell phone transmitter-receiver masts are spaced out according to the likely numbers of users as well as the ups and downs of the landscape. Masts are closer together in built-up areas.

SATELLITE NAVIGATION SYSTEM

Satellite navigation describes the use of a GPS receiver to pinpoint or "fix" your location and navigate your way from place to place. GPS is the Global Positioning System. It is a network of about 30 satellites 14,000 miles (22,530 km) high in space, orbiting Earth twice daily—plus the radio links, computers, and other equipment that control and coordinate them. A GPS receiver detects signals from the satellites and displays its own position or location on a screen.

Eureka!

The first global navigation system was Transit in the 1960s. It had five satellites, took two minutes to make a fix, and had a best accuracy of about 300 feet (91.4 m).

What next?

The United States developed the main GPS we use today, but other nations and groups are building their own systems. These include China's Compass, Russia's Glosnass, and India's IRNSS. The European Union, in partnership with several other countries, is developing the Galileo GPS.

In some cars with tinted windows and heated windshields, the metal in the tints and heating elements can prevent satellite radio signals from reaching the GPS unit inside.

Changes high in the Earth's atmosphere, especially in the layer called the ionosphere (from 30 to 620 mi or 48 to 1,000 km) can bend or weaken GPS satellite signals. Making corrections for this problem is a big challenge for GPS designers.

Batteries

Receiver The receiver unit compares the available satellite signals and locks on to the three strongest or most suitable signals.

Components Among the many chips inside, SIDRAM (synchronous dynamic random access memory) holds temporary information that can be changed quickly. It is linked or synched to the other components so that it supplies its information almost instantly on demand. In a GPS receiver, fractions of thousandths of a second are very important.

✳ How does GPS work?

At any place on Earth's surface, a GPS receiver can "see" three or more of the system's satellites, high above in space, and receive signals from them. Each satellite continuously transmits its own identity and the exact time. Radio waves are fast, but there is a slight time difference between them in reaching the receiver, because they are different distances from it. The receiver works out the delay in time from each satellite and compares them to calculate its distance from them and so work out its position on Earth's surface.

Each satellite transmits its identity and position

Some satellites are farther away than others, so signals arrive at the GPS unit at different times

GPS unit is tuned into the signals from the satellites

Read more information and watch an animation about GPS by visiting www.factsforprojects.com and clicking the web links.

Antenna (aerial) As well as a built-in antenna, the wires that connect the unit to an electricity supply may also be used to receive radio signals. The circuits are designed to detect radio signals from all available GPS satellites and to minimize the signals from other sources such as television, radio, and cell phone networks.

DGPS is used to map a glacier and follow global warming

✳ NEVER LOST AGAIN

As the name global suggests, GPS works anywhere in the world—even in frozen polar wastes or on remote mountaintops, where there is no cell phone network. Its accuracy depends partly on the cost of the receiver, with its complex circuits to compare satellite signals, but is usually within a few yards. Even better accuracy is obtained from a DGPS (differential GPS) receiver that can also detect radio signals from one or more transmitters on land.

Antiglare touch screen

Speaker Most GPS units can use prerecorded words to give instructions, such as: "At the next road junction, turn left." On many receivers, there is a choice of voices and accents.

Case

Screen The LCD color display can show various images, often a road map of the area showing direction of travel with the receiver at the lower center.

The main GPS was developed by the United States for military use, with secretly coded satellite signals. It was made available to anyone by altering the codes following an aircraft disaster in 1983 when 269 people died.

Turn In 15 y/d

Menu

GLOSSARY

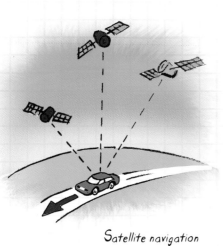

Satellite navigation system

Alloy

A combination of metals, or metals and other substances, for special purposes such as great strength, extreme lightness, resistance to high temperatures, or all of these.

Analog

Signals that carry information or data by varying in strength or size, such as by the varying strength (voltage) of an electric current.

Antenna

Part of a communications system that sends out and/or receives radio signals, microwaves, or similar waves. Most antenna, also called aerials, are either long and thin like wires, bar-shaped, or dish-shaped like a bowl.

Atmosphere

The layer of gases (air) around a very large space object such as planet Earth.

Bluetooth

A wireless system for using radio waves to communicate and send information over relatively short distances.

Blu-ray

A type of optical (light-based) disc, similar to a DVD, which uses blue laser light. This has shorter waves compared to the usual red DVD laser and so it can read smaller pits on the disc. This enables a Blu-ray disc to hold more information than a standard DVD—up to 50 GB.

CCD

Charge-coupled device, a microchip that turns patterns of light rays into patterns of electronic signals.

CD

Compact disc, a plastic-based disc with a very thin metal layer. This layer stores information or data in the form of microscopic pits, which are detected or read by a laser beam. Most CDs have a memory capacity of 650–750 MB.

Digital

Carrying information or data in the form of coded on/off signals, usually millions every second.

DVD

Digital versatile disc (sometimes digital video disc), a plastic-based disc, with a very thin metal layer. This stores information or data in the form of microscopic pits, which are detected or read by a laser beam. Most DVDs have a memory capacity of 4.7 GB or 4,700 MB. Double-sided dual-layer versions can hold more than 17 GB.

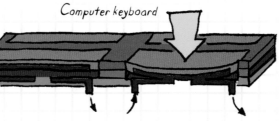

Computer keyboard

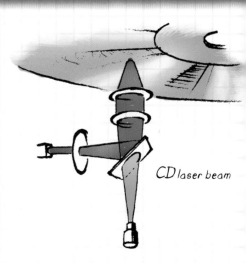

CD laser beam

Electrode

An electrical contact, or part of an electrical pathway or circuit, that is either positive (+) or negative (−).

Electron

Tiny particles inside atoms with a negative electrical charge or force that move around the central area (nucleus) of the atom. Many electrons moving along between atoms make an electric current.

Frequency

How often something happens over a certain unit of time, usually one second. For example, low-frequency sound waves vibrate about 50–100 times each second.

Gears

Toothed wheels or sprockets that fit or mesh together so that one turns the other. If they are connected by a chain or belt with holes where the teeth fit, they are generally called sprockets. Gears are used to change turning speed and force and to alter turning direction.

Giga-

One thousand million, as in 1 GB (gigabyte), which is 1,000,000,000 bytes of information or data. For a typical MP3 digital music player, 1 GB is about 16 hours of sound.

Gyroscope

A device that maintains its position and resists being moved or tilted because of its movement energy. It can be used to measure speed and direction of movement. It usually consists of a very fast-spinning ball or wheel in a frame.

HD

High definition in screen technology occurs when the tiny colored spots or pixels are smaller and closer than on a normal screen, and show more fine detail.

Infrared

A form of energy, as rays or waves, which is similar to light but has longer waves with a heating effect.

Kilo-

One thousand, as in 1 KB (kilobyte), which is 1,000 bytes of information or data. For a typical MP3 digital music player, 1 KB is about one-sixteenth of one second of sound.

Laser

A special high-energy form of light that is only one pure color. All of its waves are exactly the same length, and they are parallel to each other rather than spreading out as in normal light.

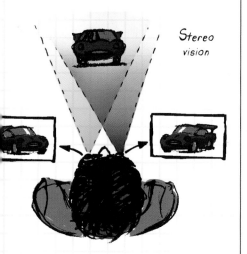

Stereo vision

Microphone

LCD

Liquid crystal display, a technology for showing images on screen using semi-liquid crystals that twist or polarize light rays.

Lens

A shaped piece of clear glass, plastic, or a similar transparent substance that bends or refracts light rays. It changes the direction of the rays so they either come together and converge, or spread apart and diverge.

Mega-

One million, as in 1 MB (megabyte), which is 1,000,000 bytes of information or data. For a typical MP3 digital music player, 1 MB is about one minute of sound.

Microchip

A small sliver or chip of a substance, usually silicon or germanium. Many microscopic electronic components and devices, such as resistors and transistors, are built into it.

Optical

Working with light rays, instead of other waves or rays (such as radio waves, microwaves or infrared heat rays).

Orbit

A curved path around a larger object, such as a satellite going around Earth, or Earth orbiting around the sun.

Pixels

Picture elements, the tiny spots or areas of different colors and brightness that make up a bigger picture or image.

Plasma

A substance similar to a gas. Its tiniest particles (atoms) get extremely hot and have an electric charge and lots of energy.

Radio waves

Invisible waves of combined electricity and magnetism. Each wave is quite long, from less than an inch to many miles. (Light waves are similar but much shorter in length.)

Satellite

Any object that goes around or orbits another. For example, the moon is a natural satellite of the Earth. The term is used especially for artificial or manmade orbiting objects, particularly those going around Earth.

Satnav

Satellite navigation, using radio signals from the GPS (Global Positioning System) satellites in space to find your way and location.

Solar

To do with the sun.

Solar panel

A device that turns sunlight directly into electricity. It is used in many small gadgets and also on satellites.

Wireless

Communicating or transferring information without wires, usually by waves such as radio, microwaves, or infrared rays.

INDEX